INSECTS AND OTHER TERRESTRIAL ARTHROPODS: BIOLOGY,
CHEMISTRY AND BEHAVIOR

PHEROMONE APPLICATIONS
IN MAIZE PEST CONTROL

INSECTS AND OTHER TERRESTRIAL ARTHROPODS: BIOLOGY, CHEMISTRY AND BEHAVIOR

Additional books in this series can be found on Nova's website under the Series tab.

Additional E-books in this series can be found on Nova's website under the E-books tab.

INSECTS AND OTHER TERRESTRIAL ARTHROPODS:
BIOLOGY, CHEMISTRY AND BEHAVIOR

PHEROMONE APPLICATIONS IN MAIZE PEST CONTROL

RENATA BAŽOK
AND
JASMINKA IGRC BARČIĆ

Novinka
Nova Science Publishers, Inc.
New York

LIBRARY OF CONGRESS CATALOGING-IN-PUBLICATION DATA

Bažok, Renata.
Pheromone applications in maize pest control / authors: Renata Bažok, and Jasminka Igrc Barcic.
p. cm.
Includes index.
ISBN 978-1-61728-010-8 (softcover)
1. Corn--Diseases and pests--Biological control--Balkan Peninsula--Case studies. 2. Wireworms--Biological control--Balkan Peninsula--Case studies. 3. Western corn rootworm--Biological control--Balkan Peninsula--Case studies. 4. European corn borer--Biological control--Balkan Peninsula—Case studies. 5. Pheromones. I. Barcic, Jasminka Igrc. II. Title. SB608.M2B39 2010 633.1'596--dc222010016716

Published by Nova Science Publishers, Inc. † New York

CONTENTS

PREFACE

In this chapter we intend to show that the potential use of the pheromones (scouting, monitoring, or optimal timing of insecticide application) depends on both the pest and the pheromone. The case of the three most important corn pests for the region of South Eastern Europe: wireworms, western corn rootworm (WCR), *Diabrotica virgifera virgifera* LeConte, and European corn borer (ECB) *Ostrinia nubilalis* Hubn. will be discussed.

Before any decision on control measures against wireworms, possible damages should be predicted. By using pheromone traps, the eight most important European species of the genus *Agriotes* in western part of Croatia five species were found. The distribution and population level of these species was established on 2 fields in central Croatia between 2001-2005. Beside pheromones on the same fields, larvae were sampled by soil sampling and by bait traps. The use of pheromone traps for wireworm pest management is attractive because of the clarity of the data they yield. There are several possibilities to use pheromones in developing IPM against wireworms: (1) monitoring and investigation in wireworm biology and ecology; (2) use of pheromones as decision tool for predicting damages and necessity for the control; (3) distribution modeling; and (4) trapping of males as a control option. There are numerous options, but all of them require further work before consistent interpretation of results could be possible.

WCR is an invasive species found for the first time in Europe in 1992. The first finding of sex pheromone of WCR dated in 1973. After the first finding the pheromone lure was produced by European scientists and pheromone trap for monitoring purposes was designed. This trap was used in Croatia for monitoring in the period between 1996 and 2006. Each year the traps were set up in corn fields (between 30-140 fields/year). Together with pheromone the yellow sticky traps (Pherocon AM or Multigard) were installed. Pheromone traps showed to be more suitable for monitoring and prediction of population increase, but, for the scouting purposes, yellow sticky traps showed to be more acceptable.

The investigation carried out in Croatia from 2002 to 2004 showed that the time of the flight of the ECB in Croatia was very variable and has changed considerably since previous years. The type E of commercial pheromone lure showed higher attractiveness than Z or E/Z types in North West Croatia. In North Croatia, E/Z and E pheromone types showed to be more attractive than Z type. Pheromones should not be used to determine intensity of the infestation, but to set the period of the maximum incidence of the moth, on the basis of which information the period of the application of insecticides is set.

INTRODUCTION

Pest management continues to be an important challenge for agricultural producers. The past few decades have witnessed a general acceptance of the necessity for considering ecology in developing pest management systems. Growers are constantly faced with the dilemma of producing a high quality pest free crop within economical means, without endangering the environment and the workers safety. Therefore, integrated pest management (IPM) was developed 50 years ago as a system approach that provides an ecologically based solution to pest control problems. IPM is defined as a sustainable approach to managing pests that combines biological, cultural, physical, and chemical tools in a way that minimizes economic, health and environmental risk. According to Cuperus, et al. (2000), the original philosophy of IPM attempted to develop a system based upon a fundamental understanding of plant/pest interactions that maintained pest populations at sub-economic levels. Later on, Kogan (1998) suggested three levels of integration: single pest management approach, integration of multiple species and methods for their management in a crop and integration of multiple species within the context of total cropping system. Pest monitoring, scouting and detection of proper application time are very important elements of the IPM. The basic tools for developing IPM systems include: agricultural control, mechanical control, host plant resistance, biological control, reproductive manipulation, chemical and regulatory control.

Monitoring in insect pest management can be used to determine the geographical distribution of pests or to asses the effectiveness of control measures. But in its widest sense monitoring is the process of measuring the variables required for the development and use of forecasts to predict pest outbreaks (Conway, 1984). A great deal of research effort is directed at developing sampling techniques and devices, particularly insect traps that may be used for monitoring (Dent, 2000).

Since the first insect sex pheromone, bombykol from silkworm moth (*Bombyx mori*) was found (Butenandt et al., 1959), the pheromones of large number of insect species have been discovered. Sex pheromones belong to the group of semiochemicals. According to Suckling and Karg (2000), semiochemicals are the odors that trigger specific behavioral response in the organism. Over the last few decades, extensive research on insect pheromones has resulted in the chemical and/or behavioral elucidation of pheromone components well over 3000 insect species, with much of the work concentrating on sex pheromones of economically important pests (Blomquist et al., 2005). There are several approaches in which pheromones can be used in pest management for the purpose of monitoring, scouting, optimal timing or for control. Monitoring the number of insects caught is the most widespread use of pheromones and there are many different ways in which this information can be used. Flight activity can be recorded as the basis for timing of insecticide application or other control tactics. Trapping can be used for efficiently monitoring the frequency or dispersion of insects or even their population traits such as insecticide resistance and for the detection of low pest densities, for example, in biosecurity quarantine programs. Pheromone lures can also provide the basis for various direct control options. The group of direct control options using attractants includes mass trapping, "attract and kill," "attract and infest," and mating disruption tactics.

Maize is one of the most important field crops worldwide. In Europe it is sown on almost 14 millions of ha (FAOstat). Maize is usually attacked by a range of different pests, but the main corn pests in Europe as well as in North America are wireworms (family Elateridae), western corn rootworm (WCR) (*Diabrotica virgifera virgifera* LeConte) and European corn borer (ECB) (*Ostrinia nubilalis* Hubn.). In order to conduct successful control of the mentioned pests farmers are using a large amount of insecticides. IPM is widely adopted in European agricultural practice, mainly in fruit production and in vegetables. It is still not utilized in filed crop production at larger scale. Therefore there is a space and need to intensify the implementation of integrated pest management (IPM) in maize production.

In this chapter we intend to show that the potential use of the pheromones in corn pest's management depends on both: the pest and the pheromone. The case of the three most important corn pests for the region of south east Europe: wireworms (family Elateriadae), western corn rootworm (*Diabrotica virgifera*

virgifera LeConte), and European corn borer (*Ostrinia nubilalis* Hubn.) will be described.

Chapter 1

CASE STUDY: WIREWORMS AND POTENTIAL USE OF PHEROMONES

1.1. INTRODUCTION

The harmful species of wireworms in Croatia belong to the genus *Agriotes*. The four main *Agriotes* species in Croatia have been described in the literature. The decision for the wireworm control should be made prior to the sowing. Therefore the key of the success in wireworm control is to have a good decision tool for the prediction of damages and the need for control measure. The knowledge on *Agriotes* species in Croatia until 2000 was limited on the data available on key species, harmfulness on different crops and the possibilities of the control (Kovačević, 1960; Maceljski & Bedeković, 1962; Maceljski, 1975). The same literature reported that the most abundant species in eastern regions of Croatia are *Agriotes ustulatus* Schall. and *Agriotes sputator* L. The species *Agriotes lineatus* L. together with *Agriotes obscurus* L. were reported as the most abundant species in western regions. *Agriotes brevis* Cand. was not mentioned. For a number of years the Faculty of Agriculture, Department for Agricultural Zoology has been conducting research on possibilities of controlling wireworms, but fundamental research on the click beetle fauna has not been carried out. A lot of research was conducted in Vojvodina and Serbia (Čamprag, 1997) and some research was conducted in the eastern part of Croatia by Štrbac (1983). But, systematic research was missing. Considering the lack of basic data on the fauna, the control is not modified to the predominant species on each particular field (or region). It is known that harmfulness varies from species to species. Species also differ in their biology and habits. These facts could have an impact on the control method.

The first results on the isolation of *Agriotes* sex pheromone were published by Yatsnin et al. (1980). It was geranyl-isovalerate isolated from *A. litigiosus*. Later on sex pheromones from females of other species were isolated: Geranyl butonate from *A. sputator* (Yatsinin et al.,1986); geranyl hexanoate and geranyl octanoate from *A. obscurus* (Borg-Karlson et al., 1988); E,E- farnesyl-acetate from *A. ustulatus* (Kudryatsev et al., 1993); geranyl octanoate and geranyl butirate from *A. lineatus* (Yatsinin et al., 1996); geranyl hexanoate from *A. sordidus/rufipalpis* (Toth et al., 2001) and geranyl butyrate and E,E, farnesyl buyrate from *A. brevis* (Toth et al., 2002). In ex SSSR different authors (Pristavko, 1988; Balkov, 1988; Balkov & Ismailov, 1991, cit. Čamprag, 1997) investigated the possibilities of the pheromones for the forecast purpose. The main species they studied were *A. gurgistanus*, *A. reiteri*, *A. lineatus* and *A. sputator*. For the first time Furlan et al. (1996, 1997) suggested the potential suitability of sex pheromone traps for implementing IPM strategies against *Agriotes* populations. Subsequently, Furlan et al. (1999) reported on the efficacy of the new *Agriotes* sex pheromone traps in detecting different species and populations. The first practical implications of the use of the new traps in Italy were described by Furlan et al. (2001). Numerous authors reported on the first results of monitoring click beetles conducted in different European countries (Toth et al., 2001a; Karabatsas et al., 2001; Gomboc et al., 2001; Furlan et al., 2001a). The main conclusion raised from all these investigations, was that the sex pheromone traps were confirmed to be a sensitive tool for detecting the key wireworm species present in one area. These traps proved to be a much more sensitive tool than soil sampling and bait traps for larvae. For all investigated species (8 species from the genus *Agriotes*) traps were able to detect wireworm populations below those that can be reliably detected using soil sampling and bait trapping. Yatsinin & Rubanova (2001) reported on the work on possibilities to use sex pheromone traps for control of wireworms by mating disruption of males. They found that there are some components which could be added to synthetic pheromones to obtain synergistic effects and improve the effect of mating disruption or mass trapping.

The studies conducted in Croatia aimed at: (1) identifying the key species in the Central Croatia, (2) evaluating the seasonal abundance of adults and larvae of *Agriotes* species in Central Croatia, and (3) assessing the effectiveness of new sex pheromone traps regarding the soil sampling and bait trapping of larvae.

1.2. MATERIALS AND METHODS

Pheromone traps for the most important European species of the genus *Agriotes* were used for monitoring Individual YALTOR funnel traps (Furlan et al., 2001a) were baited with the synthetic sex pheromones for one of the *Agriotes* species (Toth et al., 1997; Toth et al., 1998). Pheromone traps for seven most important species were set up in two fields in North West Croatia in 2001 and 2002. In 2003, 2004 and 2005 pheromone traps for five most abundant species were set up in the same fields (Oborovo and Čazma). Seven most important European species involved into research in 2001 and 2002 were *A. lineatus, A. sputator, A. obscurus, A. brevis, A. ustulatus, A. sordidus* and *A. rufipalpis.* In the period between 2003- 2005 five species, *A. lineatus, A. sputator, A. obscurus, A. brevis* and *A. ustulatus* were monitored. Both fields are situated in North West Croatia. The yearly average temperature for this region: 11.7˚C (2001); 12.3˚C (2002); 11.8˚C (2003); 11.2˚C (2004); 10.7˚C (2005), total rainfall: 822.3 mm (2001); 979.8 mm (2002); 594.1 mm (2003); 918.4 mm (2004); 906.0 mm (2005). The latitude of the locality Oborovo is 45˚41'29.25" N and 16˚15'55.84" E. The latitude of the locality Čazma is 45˚46'19.47" N and 16˚33'42.34" E. The predominant soil type in Oborovo is pseudogley soil, while in Čazma it is gley soil. The fields in Oborovo were planted with the maize or soybean (depending on year), while the fields in Čazma were planted with potato each year. The distance between the traps for different species was at least 30 m. The monitoring period for *Agriotes brevis, A. sputator, A.lineatus* and *A. obscurus* varied depending on the year and locality, but it was between April 1st and August 31st. The monitoring period for *A. ustulatus* was between April 15th and August 1st. The traps were inspected once a week. All beetle specimens were removed from the traps at each observation. Every 30 days pheromone caps were replaced. Adult population density on observed localities was classified according to Furlan et al. (2001a) as follows: High= more than 500 adults/ trap/season; Medium= between 50 and 500 adults/trap; Low= less than 50 adults /trap /season; NO= no specimens;

Based on the total individual number of five species and the individual number of each particular species, the dominance was calculated for each field and year. The dominance was calculated with Balogh's formula (cit. Balarin, 1974). The results (eudominant, dominant, subdominant, recedent, subrecedent) were classified according to Tischler and Heydeman (cit. Balarin, 1974).

In order to establish the presence of the larvae in the period between 2001 and 2005, field soil bait traps and soil samplings were made. Fifty soil bait traps were placed each year in autumn and spring in a grid 20 m x 30 m covering the area where the pheromone traps were set up. Soil bait traps were made and used according to the description given by Chabert and Blot (1992). The traps were placed in the fields in March and in October. After 15-20 days they were removed and checked by hand-sorting the contents. After hand sorting bait traps were processed in Tullgren funnels. Soil samples were taken 1-2 m away from the soil bait. They were 11 cm in diameter and 30 cm deep (covering 0.0095 sqm). After the collection soil samples were processed by putting soil cores into Tullgren funnels.

1.3. RESULTS AND DISCUSSION

By using pheromone traps for the eight most important Europen species of the genus *Agriotes,* in the north western part of Croatia, five species were found. *Agriotes sordidus and A. rufipalpis* were not captured at any locality in 2001 and 2002. Therefore, the pheromones for those 2 species were not operated in 2003-2005.

The species *A. lineatus* was found in high population at both localities and in all the years except at locality Oborovo in 2005. The swarming period for this species started early in spring (April) and lasted through the whole period of monitoring until mid of August (Figure 1). The similar results were obtained in Hungary (Toth et al., 2001a), in Slovenia (Gomboc et al., 2001) and in Netherland (Ester at al., 2001).Out of 22 weeks of monitoring adults were captured in 14 to 19 weeks. The peak of flight was clearly visible only in few cases. At locality Oborovo in 2001 and in 2004 it was the end of May what is similar to the results obtained in Hungary (Toth et al., 2001a), Slovenia (Gomboc et al., and Netherland (Ester et al., 2001). At locality Čazma in 2004 the peak of appearance was in August. Generally, adults of *A. lineatus* could be captured during the whole vegetation season, between April and August. It corresponds with the data on the biology of this species. *A. lineatus* over-winters as adult and are active during the whole season. Maximal weekly captures of adults of *A. lineatus* in Croatia in some years were very high. They exceeded 700 beetles/ trap/ week in Oborovo in 2001 and in Čazma in 2002 and 2004. In Hungary and in Slovenia maximal weekly captures reached 100 and 140 beetles per week respectively (Toth et al., 2001a; Gomboc et al., 2001). Comparing the investigated localities somewhat higher population was established at locality Čazma. In three, out of five years of investigation, total capture of *A. lineatus* exceeded 3,000 beetles. The appearance of *A. lineatus* in Croatian literature was always related to humid and cold climatic conditions (Maceljski, 2001). However, warmer soil conditions are usually recorded at locality with higher capture (Čazma) than at locality Oborovo and this statement could be discussed.

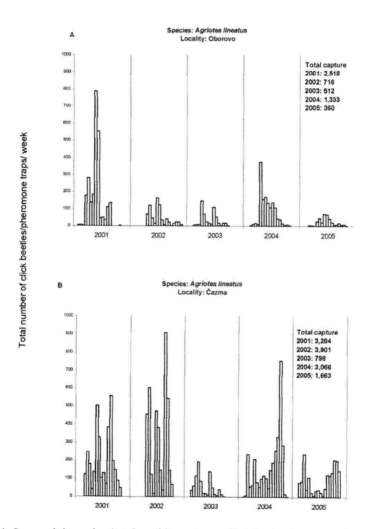

Figure 1. Seasonal dynamics (total weekly captures of beetles in pheromone traps between April 1st and August 31st) of the species *Agriotes lineatus* observed in years 2001-2005 at locality Oborovo (A) and Čazma (B).

The literature indicated that in Croatia the species *A. obscurus* usually appears at the same areas as *A. lineatus* (Maceljski, 2001). Since the adults of *A.obscurus* were caught at both localities in low to medium population, the results are confirming this statement. But, comparing to *A. lineatus* the total number of captured adults (Figure 2) was low. At locality Oborovo they varied between 29 and 123, and at locality Čazma between 40 and 79 beetles. Toth et al. (2001a) did not find this species in pheromone traps even though in literature it was mentioned as one of the most dangerous for Hungary. The swarming period depended on year and locality. It started as earliest in the 3rd decade of April and lasted as longest until the end of July (Figure 2). The swarming period for this species in Slovenia (Gomboc et al., 2001) is similar, while in Netherland (Ester at al., 2001) the flight started at the end of May and lasted short, till the end of June. Out of 22 weeks of monitoring adults were captured in 7 to 16 weeks. The peak of flight was at the end of May and the beginning of June, as it is in Slovenia (Gomboc et al. 2001). The adults of *A. obscurus* start to appear in the field early in the spring. It confirms that *A. obscurus* over-winters as adult in soil. Even though the literature correlated the appearance of this species with *A. lineatus,* the correlation in population level between those two species did not exist in this investigation. The calculated correlation coefficient was very low and not significant. Gomboc et al. (2001) and Ester et al. (2001) investigated both species in Slovenia and in Netherlands, respectively. Their results indicate stronger correlations in population level between those two species. In both investigations *A. lineatus* was captured in higher population level than *A. obscurus*. In some parts of Europe (Furlan et al., 2001a) population of *A. obscurus* was higher than *A. lineatus.* The difference probably results in different climatic and edaphic conditions.

The capture of *A. sputator* ranged from low (27 click beetles at Oborovo in 2002) to medium (498 at Čazma in 2001), depending on the year and locality. The swarming period started early in spring (mid April) and lasted as longest until the end of July (Figure 3). Toth et al. (2001a) showed similar results for Hungary and they indicated the strong impact of locality and year on the obtained results. The swarming periods for this species in Slovenia (Gomboc et al., 2001) and in Hungary (Toth et al., 2001a) are similar with those noted in Croatia. Out of 22 weeks of monitoring adults were captured in 5 to 17 weeks. The peak of flight was recorded in May. The adults of *A. sputator* start to appear in the field early in the spring, earlier than *A. obscurus* and *A. lineatus*. It confirms that *A. sputator* over-winters as adult in soil. Larvae of *A. sputator* together with *A. lineatus* and *A. obscurus* are recognized in the UK as wireworms and are particularly important to potato growers (Hicks & Blackshaw, 2008).

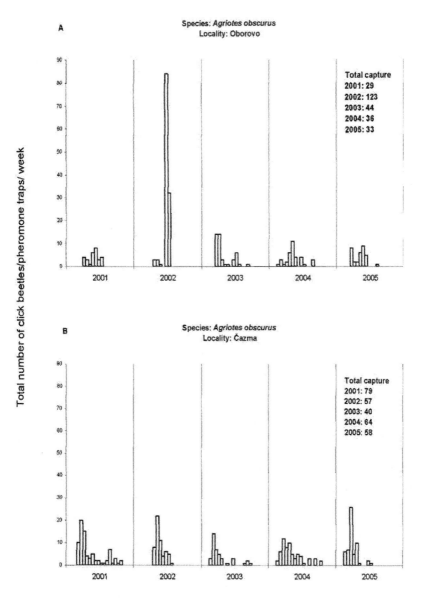

Figure 2. Seasonal dynamics (total weekly captures of beetles in pheromone traps between April 1st and August 31st) of the species *Agriotes obscurus* observed in years 2001-2005 at locality Oborovo (A) and Čazma (B).

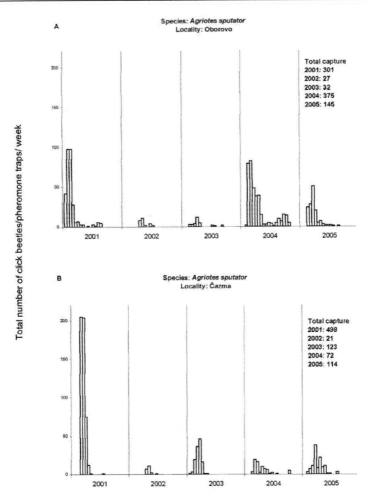

Figure 3. Seasonal dynamics (total weekly captures of beetles in pheromone traps between April 1st and August 31st) of the species *Agriotes sputator* observed in years 2001-2005 at locality Oborovo (A) and Čazma (B).

The species *A. brevis* has never been reported as an important species in Croatian agricultural areas. It was recorded by Novak (1952) as one of the species of Elateridae family in the coastal area of Croatia. The present investigation shows that *A. brevis* is present at medium to high population at both localities (Figure 4). This species has been found in Slovenia (Gomboc et al., 2001), Bulgaria and Romania in low to medium population (Furlan et al., 2001a). Furlan et al. (2001) recorded medium to high population in northern regions of Italy, while in other European countries it has not been found

(Furlan et al., 2001a). Morphologically, this species is difficult to distinguish from *A. sputator*. Similarities between these two species are not only in morphological characteristics, they are present in biological characteristics as well as in their reaction to pheromone baits. Those two species are systematically very close. Our recent work (unpublished) shows that very few (less than 5%) of *A. sputator* are captured by the pheromone for *A. brevis*. The swarming period of *A. brevis* was similar to those of *A. sputator*. The appearance of adults started with the beginning of April. The peak of flight was in May.

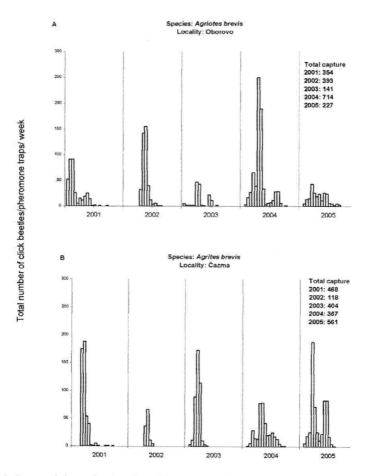

Figure 4. Seasonal dynamics (total weekly captures of beetles in pheromone traps between April 1st and August 31st) of the species *Agriotes brevis* observed in years 2001-2005 at locality Oborovo (A) and Čazma (B).

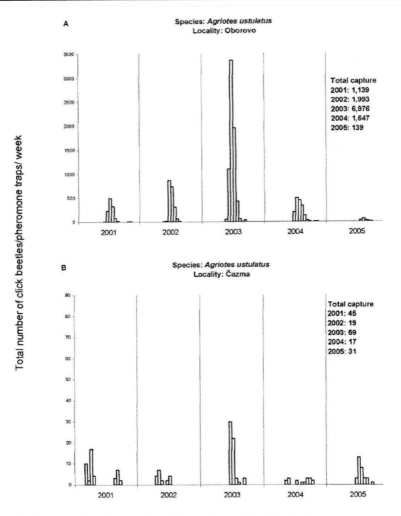

Figure 5. Seasonal dynamics (total weekly captures of beetles in pheromone traps between April 1st and August 31st) of the species *Agriotes ustulatus* observed in years 2001-2005 at locality Oborovo (A) and Čazma (B).

One of the most dangerous species is ***A. ustulatus***. Among all monitored species, this species has the shortest life cycle with larvae developing in soil for 2-3 years (Furlan, 1998). It was mentioned that this species was dominant in the eastern part of Croatia (Kovačević, 1960; Maceljski, 2001). The high population established at locality Oborovo (Figure 5) shows that *A. ustulatus* has changed its range of spreading. Kozina et al. (2008) reported that the dominance of *A. ustulatus* has increased in the north western region and now it is the most dominant species in all regions of Croatia. Low to medium population of *A. ustulatus* was reported by Gomboc et al. (2001) in Slovenia and Furlan el al. (2001a) in Germany and Romania, medium population was reported by Toth et al. (2001a) in Hungary and medium to high population was reported by Furlan et al. (2001) in some parts of north Italy. The swarming period of *A. ustulatus* is shorter comparing with other species. It lasted up to 8 weeks with the peak of flight at the end of June.

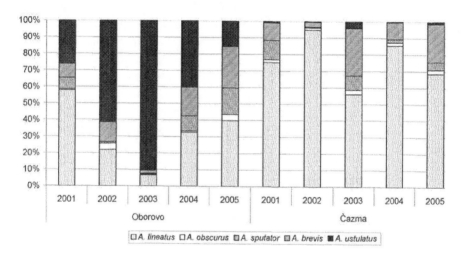

Figure 6 The ratio of five *Agriotes* click beetles on pheromone traps at two localities in Croatia.

The dominance index of *A. lineatus* at locality Oborovo (Figure 6) ranged between 7% (in 2003) and 58% (in 2001). The same species was eudominant at locality Čazma with dominance index between 56% (in 2003) and 95% (in 2002). *A. obscurus* was subdominant, recedent and subrecedent at both localities with dominance index ranged between 0.6 and 4%. The dominance index of *A. sputator* varied at high extent. Depending on year and locality, the species was ranged in all five categories. The species *A. brevis* could be classified as subdominant, dominant or eudominant, depending on year and locality. The species *A. ustulatus* was eudominant at locality Oborovo with dominance index between 15 and 90%, while the same species was subdominant, recedent or subrecedent at locality Čazma. The distance between those two localities is not very big, it is approximately 30-40 km and the climatic conditions could not explain the difference. Edaphic conditions could not be the explanation for the difference because the soil in Oborovo belongs to pseudgley soils which are heavy and difficult to warm up. The soil at locality Čazma belongs to gley soils and is not so heavy. This soil is warmer comparing to Oborovo what fits better to *A. ustulatus*. At the experimental field in Oborovo potato has been planted for five years. Čamprag (1997) stated that potato is not very suitable for larvae of *A. ustulatus* to complete their life cycle what could be a possible explanation for low population of *A. ustulatus* on this field.

Although the population level of *A. lineatus* in pheromone traps was high in all years and localities, larvae were found only occasionally in low numbers. Population level of *A. obscurus* was low, but larvae were found quite often (Table 1). Population level of males of *A. sputator* was low to medium and larvae were found in 4 out of 8 samples. Adult population of *A. brevis* was medium to high and larvae were found only once. Larvae of *A. ustulatus* were found in all samples from locality Oborovo where adult population was determined as high. Larvae were not found in samples from locality Čazma in period 2001-2003 where the adult population was determined as low to medium. In 2004 and 2005 larvae of *A. ustulatus* were found in samples from locality Čazma even though the adult population was low.

Blackshaw & Verrnon (2008) stated that pheromone traps, as currently used, will not reliably indicate where wireworms occur in a field, and that the complexity of interpreting adult male trap counts limits quantitative predictions of population size. According to Hicks & Blackshaw (2008) pheromones could be used as a decision tool for wireworm management in potato crops. It is recommended to use pheromones in a set of three dominant species in UK, *A. lineatus*, *A. obscurus* and *A. sputator*. If the total capture is less than 50 beetles some predicted wireworm population will be 25.000-250.000/ha and some damage could occur. If the total capture is between 50 and 100, significant damage could occur and predicted population level is 150.000-250.000/ha. The capture over 150 beetles means predicted population level over 250.000/ha (25 larvae/sqm) and severe damage. Furlan (2009) also developed similar system for *A. brevis* and *A. ustulatus* in Northern Italy. He indicated higher captures for economic threshold level. If more 300 *A. brevis* are captured in one pheromone trap in one year, it considers the high population. In the next year the average of 1 larva of the same species in soil bait will be found. He also reported that the total capture of 1000 *A. ustulatus* could result with 4 larvae/soil bait and considers high population of this species.

Table 1. Classification of adults capture (according to Furlan et al., 2001a) and result of larval sampling at localities Oborovo and Čazma, Croatia 2001-2005

		2001 Oborovo	2001 Čazma	2002 Oborovo	2002 Čazma	2003 Oborovo	2003 Čazma	2004 Oborovo	2004 Čazma	2005 Oborovo	2005 Čazma
Adult species	A. lineatus	H	H	H	H	H	H	H	H	M	H
	A. obscurus	L	M	M	M	L	L	L	M	L	M
	A. sputator	M	M	L	L	L	M	M	M	M	M
	A. brevis	M	M	M	M	M	M	H	M	M	H
	A. ustulatus	H	M	H	L	H	M	H	L	M	L
Number of larvae/ 50 soil baits		111	18	2	7	5	24	23	76	82	106
Number of larvae/50 samples		11	4	1	4	11	11	7	18	3	7
Average number of larvae/m2		23,1	8,4	2,1	8,4	23,1	23,1	14,7	37,9	6,3	14,7
Larval identification	A. lineatus	No	No	No	No	No	Yes	*	No	*	Yes
	A. obscurus	No	No	No	Yes	Yes	Yes	*	No	*	No
	A. sputator	No	Yes	No	No	Yes	No	*	Yes	*	Yes
	A. brevis	No	No	No	No	No	Yes	*	No	*	No
	A. ustulatus	Yes	No	Yes	No	Yes	No	*	Yes	*	Yes

* Larvae were not identified.

The results obtained in presented study correspond better to the data obtained by Furlan (2009) than to the results reported by Hicks & Blackshaw (2008). During the investigation in Croatia the total captures were over one or more thousand of click beetles, and larval infestation established by soil sampling was slightly over the threshold at all localities and years. The highest infestation of 37.9 larvae/ sqm was found (Table 1). The damages on experimental plots were present, but we did not experience visible damages on the stand plant counts in maize and soybean. Potato tubers were not examined for damage so it is not possible to conclude the level of damages. Probably the reasons could be in different climatic and agronomic conditions.

In Croatia, there is an outgoing project "Spatial distribution patterns of economically important pests using GIS methods." The proposed investigation is aimed at testing GIS applicability to entomological research. Within the frame of this project a research on wireworm's spatial distribution patterns are conducting. The objectives of the research of wireworms cover: A) Making maps of spatial and temporal distribution of five most harmful species of the family Elateridae; B) For each of the species investigated, the correlation of incidence and intensity and the most important agroecological factors will be determined in the same way as the elements that affect spatial and temporal distribution will be determined; and C) Everything achieved will create the basis for the modeling of the incidence and scope of an attack of each of the pests, which will be useful in forecasting attack intensity and incidence of these pests. The results of the research will enable a better understanding of the factors that affect the spatial distribution of the 5 most important species of wireworm. In conjunction with data about their harmfulness, this will give better understanding of the need for control. The results will enable a more complete and better understanding of factors that lead to population changes in these species.

Based on results on differential responses of three *Agriotes* species to pheromone traps, Hicks & Blackshaw (2008) calculated the average price for the control by trapping. They calculated that ten *A. lineatus* traps/ha and fifteen *A. obsurus* traps/ha and 142 *A. sputator* traps/ha are needed. If trapping period lasts from April to August, 100% success in reducing mating and oviposition will be achived and this will cost 2,755.5 €/ha. Trapping should be done over 4 years.

1.4. CONCLUSION

The use of pheromone traps for wireworm pest management is attractive because of the clarity of the data they yield. There are several possibilities to use pheromones in developing IPM against wireworms: (1) monitoring and investigation in wireworm biology and ecology; (2) use of pheromones as decision tool for predicting damages and necessity for the control; (3) distribution modeling; and (4) trapping of males as an control option.

Pheromone traps show to be very sensitive and thus acceptable as monitoring tool for the most dangerous *Agriotes* species present in many European countries. The investigations of the fauna of the genus *Agriotes* conducted in different European countries by using pheromone traps showed that the previously mentioned literature data does not always correspond with current situation. The traps are very selective and capable to attract the desired species. Adult male click beetles (*Agriotes* spp.) show good, but differential responses to specific pheromone traps. Therefore traps are suitable for the investigation of the flight dynamics. They could be used for studying differences in flight patterns between sites, species and years.

The question is whether pheromone traps can be used to predict the size of the wireworm population and the need for insecticide application. The problem is that in one field mixed population of wireworms exists. The trapped adults arise from a different generation to the wireworms that remain as pests in the soil; thus, for them to quantitatively reflect population levels there have to be assumptions that cohorts are similar in size with dynamics that are largely uninfluenced by environmental experience. Before any use of pheromone traps for the prediction of population size, on the same area, the prevalent (eudominant and dominant) species of click beetles and larvae should be identified. Hicks and Blackshaw (2008) agree that recommendations based on the cumulative total catch of three species over a sampling season can be improved by considering the spatial relationship between the adult trapping system and larval distribution. The current constraint to this is the general inability to separate wireworms into species.

If population levels and different flight patterns are correlated with climatic (temperatures and humidity), edaphic (soil type) and agronomic (crop and tillage type) data, more information on factors influencing spatial distribution could be available. Species distribution modeling is becoming an important tool for understanding the effects of changing environmental conditions on biogeographical patterns.

Until now, there is no experience with male trapping as control option carried out somewhere in the agricultural practice. But according to Hicks & Blackshaw (2008), this is a realistic idea which is expensive and, as we assumed, not affordable for farmers in the UK. If this method were used somewhere else, prior investigation on species composition would be needed.

In conclusion, there are numerous options but all of them require further work before consistent interpretation of results could be possible.

Chapter 2

CASE STUDY: WESTERN CORN ROOTWORM AND POTENTIAL USE OF PHEROMONES

2.1. INTRODUCTION

The western corn rootworm (WCR), *Diabrotica virgifera virgifera* LeConte, is the most serious pest of maize across the US Corn Belt (Metcalf, 1986). The pest was recently introduced into Europe where it was first observed near Belgrade, Serbia, in 1992 (Bača, 1994). Szalai et al. (personal communication) estimated that the first introduction probably has occurred between 1982 and 1984, i.e. 8 to 10 years before the first detection of this species and its damages in maize fields. Once introduced, *D. virgifera* started to spread across Europe. The first finding of sex pheromone of WCR dated in 1973 (Ball & Chaundry, 1973, cit. Guss at al., 1976). The first trials with natural pheromones collected on filter paper from females before they copulated were conducted by Guss et al. (1976). Bartlet and Chiang (1977) estimated that if 8 pieces of filter papers with pheromones were applied on one ha, 1.1 to 5.8% of all males from the same area could be captured. Successful isolation of pheromone (Guss et al., 1981) was followed by successful synthesis in laboratory conditions. The investigations under natural conditions in the USA conducted by Guss et al. (1984) and Lance (1988) proved that pheromone is more attractive to WCR than to *Diabrotica barberi* (NCR). Due to high price and availability of other suitable and cheaper monitoring tools the investigations in the USA terminated.

After the first appearance of WCR in Serbia surrounding countries started with monitoring of WCR for the detection purpose. In the first year cucurbitacin baited traps were used. The captures in the first year of monitoring were low (Igrc Barčić, 1996), due to low attractiveness of cucurbitacin and due to low level of infestation (Ilovai, 1996). At the same time, following the data from the USA, Toth et al. (1996) have synthesized pheromone and carried out the first trials in Serbia. They optimized the shape of the trap and the durability of the pheromone capsule. After 1996 all European countries started to use pheromones in their monitoring programs. Simultaneously, with the use of pheromone traps for monitoring purposes, the investigations were conducted in order to collect more information on attractiveness of pheromones (comparing with other types of attractants) and to estimate the maximum sampling range from which the pheromone trap is able to attract any male WCR beetle. Traps showed to be very efficient when the population level was low (Igrc Barčić & Dobrinčić, 1998; Igrc Barčić et al., 1999). In the conditions of high population level pheromone traps captured similar number of beetles as the yellow sticky traps did (Kereši et al., 1997; Edwards et al., 1997; Zseller & Szell, 2000). Toth et al. (2000) proved that the maximum sampling range of pheromone traps is over 100 m, but that the zone of attractiveness is shorter than 10 m.

The monitoring of WCR in central european countries started in 1996. The International Working Group on *Ostrinia* and Other Maize Pests (IWGO) of the International Organization for Biological Control (IOBC) organized the first WCR international workshop in Europe in cooperation with FAO and EPPO in March 1995 (Berger, 1996). The result of this meeting was the establishment of the IWGO *Diabrotica* Subgroup (Kiss et al., 2005). The first detections of WCR adults in Croatia (Igrc Barčić & Maceljski, 1996) and in Hungary (Prinzinger, 1996) occurred in 1995. Owing to regular contacts and agreements within the said group, WCR became the only pest in the world whose monitoring proceeded using the same methodology in most countries and whose spread was identified in detail each year. Starting with 1996 in all monitoring actions in all countries, pheromone traps were used. The monitoring of adult WCR by European countries has allowed the rapid detection and determination as to the spread of the invasive pest species since the insect was first observed in Serbia (Kiss et al., 2005). In each network partner, the permanent monitoring sites were established. The permanent monitoring sites have allowed for the measurement of population fluctuations over the years.

Between 1996 and 2006, the monitoring of WCR was regularly conducted in infested and non-infested area in Croatia. The aims of the monitoring activity were (1) to establish the spreading of WCR across Croatia, (2) to establish attractiveness of pheromone traps vs. yellow sticky plates, (3) to establish the flight dynamic of WCR adults, and (4) to establish population changes over years.

2.2. MATERIAL AND METHODS

Pheromone traps (Csalomon) were used in Croatia for monitoring of WCR in the period of 1996-2006. The monitoring was conducted in 7 counties in 1996, in 8 counties in 1997, 1998 and 1999, 11 counties in 2000, 2001 and 2002, and in 13 counties in 2003, 2004 and in 2005. In 2006 the monitoring was conducted in 11 counties. Each year traps were set up in maize fields (between 31-148 fields/year) situated in different areas of Croatia. Together with pheromone traps Pherocon® AM (PhAM) non-baited yellow sticky traps (Treece, Salinas, USA) or Multigard® (Scentry, Billings, USA) non-baited yellow sticky traps were installed.

In each field, a set of one pheromone and one yellow sticky trap was installed. The traps were placed in the fields at minimal distance of 50 m. They were placed in the fields at the end of June, every 7 days the capture of the beetles was recorded and all captured beetles were removed, pheromone capsules and yellow sticky traps were replaced by new ones every 4 weeks. The monitoring terminated at the end of September.

2.3. RESULTS AND DISCUSSION

During the 11 years of monitoring in Croatia, we were able to predict the line of spread of WCR for each year (Figure 7). WCR spread over a distance of 20 to 60 km per year toward the west. Pheromones showed to be very sensitive for early detection purposes. It is visible on the example of year 2001 in which at only one isolated locality, 25 km deep in un-infested area, the new capture was recorded (Figure 1). In 2002, the whole area between the 2000 spread line and those isolated localities in 2001 was infested by beetles. The neighboring and other European countries have conducted monitoring of WCR by using the same pheromone (Csalomon) traps. As a result of well coordinated action, the scientific community was able to follow the spread of WCR across Europe. Every year the map of spread of WCR across Europe has been published on EPPO web site: http://www.eppo.org/ QUARANTINE/ Diabrotica_virgifera/diabrotica_virgifera.htm. The data obtained through the monitoring action conducted in Croatia were incorporated into this map. The spread of WCR across Europe between 1995 and 2003 was analyzed by Kiss et al. (2005). It is interesting that WCR spread from its first point of introduction in all directions. The average speed of this so called "natural" spread was within few kilometers, as observed in Hungary in 1998, or 70-80 km, as seen in Hungary in 1999 and in Romania in 2001. Every year the new 25,000 to 65,000 sqkm of European area was infested by WCR (Kiss et al., 2005). This trend continued after 2003 as well (see the spread map at EPPO web site). Because of general awareness, all European countries started to use pheromone traps for early detection of adults. The pheromones were placed mainly near airports, railway stations etc. Beside the "natural" spread in Italy in 1998 a so-called "jumping spread" occurred for the first time. Jumping spread means the occurrence of adults far from the actual spread line (Kiss et al., 2005). Jumping spread occurred in Switzerland (Bertrossa et al., 2001) and again in Italy near Milan in 2000. The next jumping spread occurred in France in 2002 (Reynaud, 2002). In 2003 several jumping spreads occurred in Belgium, Netherland and the UK. All jumping spreads occurred near the airports and were detected by pheromone traps what proved a high sensitivity and suitability of pheromones for early detection purposes. Furlan et al. (1998) reported on the first finding of WCR near the Venice airport in Italy. It was the first case of so-called "jumping spread". In 1999, Furlan et al. proposed eradication and/or containment strategies against WCR based on defining focus and safe area, trapping of WCR males on pheromone traps, prohibiting continuous maize in focus area, applying insecticide treatments to maize fields to control WCR beetles, forbidding the movement of fresh maize or soil in which maize was grown the previous year etc. The proposed strategy resulted with effective stopping of WCR populations (Furlan et al., 2001b) in the first years of their conduction. Even though original focus area was kept free of beetles, after 8 years the program was stopped (Furlan et al., 2009). WCR populations were moving towards the focus area from new infested areas which were discovered in Italy in 2000. The sources of jumping spreads were investigated by employing genetic studies. Ciosi et al. (2009) investigated within- and between-population variation, at eight microsatellite markers, in WCR in North America and Europe, to investigate the routes of introduction of WCR into Europe and to assess the effect of introduction events on genetic variation. From 1992 to 2004, they detected five independent introduction events from the northern US to Europe and two intra-European foundations.

Figure 7. The spread of WCR in Croatia, 1995-2006.

Table 2. The capture of Western Corn Rootworm on different types of traps in monitoring conducted in Croatia in period 1996-2006

Year of monitoring	Number of installed traps		Number of fields with capture		Number of captured beetles		Ratio of captured beetles in total capture (%)	
	Ph*	Yellow trap **	Ph	Yellow trap	Ph	Yellow trap	Ph	Yellow trap
1996	119	60	44	4	769	14	98.21	1.79
1997	134	90	58	10	3,141	86	97.33	2.67
1997	136	136	69	20	5,628	305	94.86	5.14
1999	122	122	67	38	14,570	1,218	92.3	7.7
2000	130	130	81	42	14,487	597	96.04	3.96
2001	148	148	80	43	20,899	809	96.32	3.68
2002	140	140	86	44	22,704	1,610	93.38	6.62
2003	121	121	90	56	29,673	2,087	93.43	6.57
2004	59	59	38	25	20,777	3,342	86.14	13.86
2005	35	55	17	21	2683	1659	***	***
2006	31	31	25	12	11,301	1010	91.8	8.2

* Ph- Pheromone trap (Csalomon).

**Multigard –M traps were installed in period 1996-1999, and on part of the fields in 2000, and Pherocon AM (PhAM) were installed in 2000 (partially) and in period 2001-2006.

*** Because the traps were not installed on the same fields, the ratio could not be calculated.

In the first years of monitoring, Multigard yellow sticky traps were used, but in 2000 they were replaced by PhAM traps. The switch was made due to the fact that more data were available from the USA on what the actual numbers caught on PhAM traps meant in regard to the possibility of the presence of WCR economic populations. Youngman et al. (1996) reported that the higher captures occurred on Multigard than on PhAM trap. This was proved by Barna et al. (1998). The ratio of beetles captured on Multigard traps in the period of 1996-1999 varied between 1.79 and 7.7% of all captured beetles (Table 2). It was obvious that the ratio of beetles captured on Multigard traps increased with the increase of the total capture of beetles. When the data from the period between 1996-1999 were correlated (Figure 8), positive correlation ($r=0.93$) was established. The regression curve is linear. In the period 2001-2004, PhAM traps captured 3.68-13.86% of all captured beetles. When the data from the period between 2001-2004 were correlated (Figure 8), medium positive correlation ($r=0.503$) was established. Regression curve is polynomial. Anyhow pheromone traps captured over 90% of all beetles what proves its higher attractiveness for WCR males comparing to Multigard and PhAM traps. The ratio of the beetles captured on PhAM and Multigard traps depends on the choice of sites. If the majority of monitoring sites are located in infested area with established high population, the ratio of beetles captured on visual traps will be higher than if the majority of monitoring sites are located at the line of spread and in un-infested area with low population. Even some authors (Kereši et al., 1997; Edwards et al., 1997; Zseller & Szell, 2000) reported that in the conditions of high population the capture on pheromone traps and yellow sticky traps (Multigard) were the same. Our results with PhAM traps did not prove this statement. In the case of economic adult population the main goal should be to establish the population level and to predict the need for insecticide application against larvae or the need for crop rotation. For this purpose in the USA different possible scouting tools were developed (Hein & Tollefson, 1984; Wilde, 1999). Due to high attractiveness and high number of beetles which should be counted pheromones are not suitable for this purpose. In the areas with high population, the sticky traps are not suitable, because the sampling efficiency of the trap is constantly changing over time depending on the number of insects already captured and other debris accumulating on the sticky surface.

The flight dynamics is presented as the ratio of beetles captured in a certain period comparing to the total capture of the beetles. The curves of the flight dynamics (Figure 9) depended on the year. In the years with warm and dry spring (2000, 2002 and 2003) the beetles started to emerge earlier, and the peak of the flight was in July, while in other years the peak of the flight was in August. In the years with warmer conditions the flight terminated by the end of August while in the years with "regular" (average) conditions the flight terminated after the first decade of September. In some years (1997, 2004, 2005 and 2006) the flight lased up to mid October.

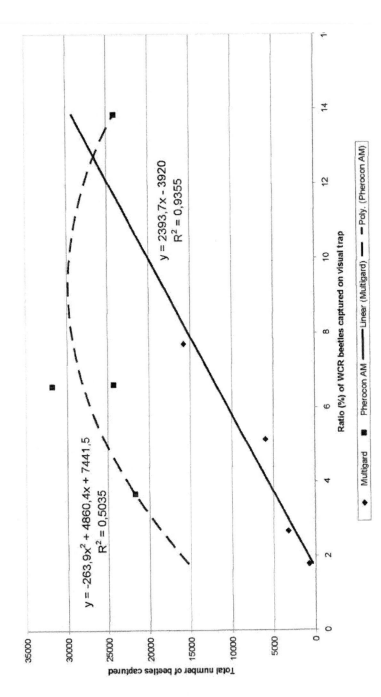

Figure 8. The relationship between ratio of beetles captured on visual traps (Multigard and Pherocon AM) vs. total number of beetles captured in given year.

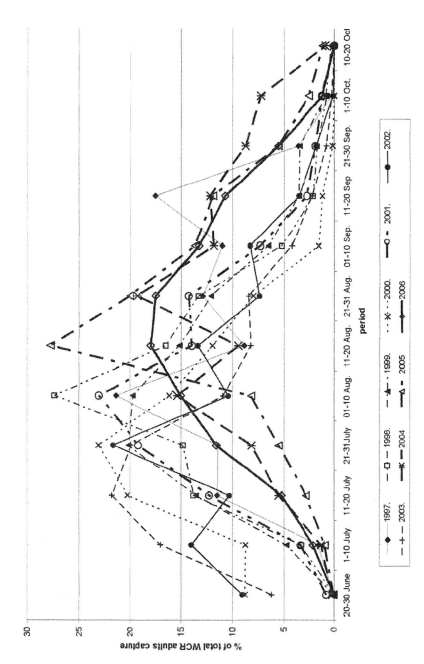

Figure 9. The dynamics of the flight of WCR in Croatia 1997-2006.

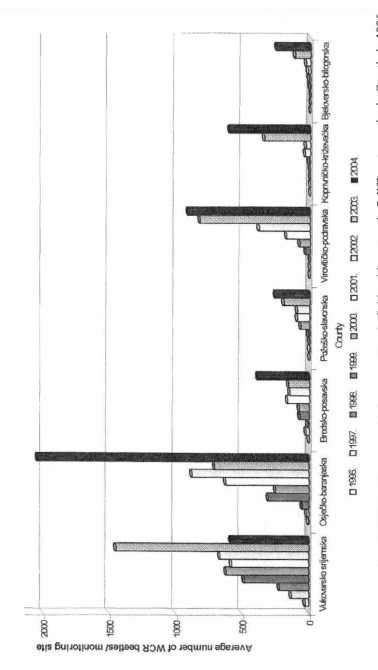

Figure 10. Average number of WCR beetles captured/ pheromone trap in fields with capture in 7 different counties in Croatia in 1996-2004.

The adult activity period reflects on the population level in subsequent year. If the adult activity as well as oviposition period is long, females could oviposit the higher number of eggs and in the subsequent year high larval attack could be expected. The flight dynamics could indicate on adult longevity under field conditions. Topfer & Kuhlman (2005) indicated that under field conditions in Central Europe the mean female longevity varied between 1 and 1.5 months. In the laboratory, the mean longevity of field collected females was approximately 3 months (Topfer & Kuhlman, 2005). Igrc Barčić & Bažok (2004) compared the life span of the field-collected WCR beetles from the USA and Croatia in laboratory conditions and concluded that there are minor differences. The life span was 63.7 and 78 days for beetles from Croatia and 87 days for the beetles from the USA respectively. But, they did not observe the differences in oviposition period between the populations of different origins. The trapping period lasted up to 4 months. It should be kept in mind that the presented flight dynamics is the flight dynamics of males (they are attracted by pheromones), and it could be somewhat different comparing to females. It is known that females start to emerge later than males, the difference in emergence is 3-4 days in Croatia (Dobrinčić, 2001), and it could be predicted that females would live a few days longer.

Monitoring was also used to identify the pest's population density in the already infested areas. The data on population expansion in the observation spots and within affected counties have allowed the measurement of population fluctuations over the years (Figure 10). Based on the average number of beetles/monitoring site with capture in different regions, the average index of population growth for each county was calculated. Occasionally, it would expand 2- 6 times annually. It has to be shown that WCR populations in one county may increase by two to six times when compared to previous year, but they may decrease as well. Decreases are most probably the effect of the cumulative consequences of unfavorable weather conditions for WCR, such as growing maize in crop rotation with non-host crops. This was also proved by the result of other European countries (Kiss et al., 2005).

Based on the established lines of spread the infested area for each year was calculated. Based on the data from Statistical Annual (1996-2007), the acreage of maize in the infested area was predicted (Table 3). Based on the data collected from Pherocon AM at monitoring sites in the infested area, the area of economic adult activity was predicted as the acreage of maize in this area (Table 2). The infested area of Croatia in 2006 was 25,900 sqkm (the total area of Croatia is cca 56,000 sqkm). In this area approximately 300.000 ha of maize is sown. It represents over 90% of all maize acreage in Croatia. Starting form 1995, every year the infested area increased up to 5,000 sqkm. The data for Europe showed that every year the new 25,000 to 65,000 sqkm of European area was infested by WCR (Kiss et al., 2005). The area of economic adult activity (area in which WCR adult population is high enough to cause economic larval damage to maize in subsequent year) covered 9,400 sqkm. In this area approximately 118,000 ha of maize is sown out of which 25% is continuous maize. The total area of WCR economic adult activity in Europe in 2003 was 70,500 sqkm (Kiss et al., 2005). It was distributed in Hungary, Romania, Serbia, and Croatia.

Table 3. Spread of WCR and adult activity area in Croatia, 1995-2006

YEAR	INFESTED AREA (sqkm)	Acreage of maize in infested area (ha)	Area of economic adult activity* (sqkm)	Acreage of maize in economic adult activity (ha
1995	1,200	10,000	-	-
1996	6,500	85,000	-	-
1997	9,000	125,000	-	-
1998	10,000	140,000	-	-
1999	12,750	165,000	-	-
2000	14,000	200,000	2	100
2001	15,500	220,000	4,000	30,000
2002	19,000	250,000	5,000	50,000
2003	23,500	290,000	7,500	95,000
2004	24,500	328,000	7,500	95,000
2005	24,500	328,000	8,500	105,000
2006	25,900	290,000	9,400	118,000

* Economic adult activity refers to WCR adult population that can result in economic larval damage to maize in subsequent year.

2.4. CONCLUSION

Conducted research of pheromones for WCR proved that in the case of WCR they are suitable for certain purposes only. Among the previously mentioned approaches in which pheromones can be used in pest management, the use of pheromones for WCR should be oriented in two directions: because of their high sensitivity they are suitable for early detection purposes in biosecurity quarantine programs and for containment strategies or they can be used to establish population changes over years.

The case of introduction of WCR into Europe and all work conducted with monitoring purposes in Europe proved that by using pheromones very precise detection of the spread line of WCR in all European countries was possible. Furlan et al. (2009) proved that the use of pheromones together with some other well planned measures could be effective in containment strategies if an isolated introduction is found. Pheromones are suitable to measure population fluctuation over years, but if the economic adult population is present, yellow sticky traps should be employed.

Chapter 3

CASE STUDY: EUROPEAN CORN BORER AND POTENTIAL USE OF PHEROMONES

3.1. INTRODUCTION

The European corn borer (ECB) (*Ostrinia nubilalis* Hübner) is the most important pest in Croatian agriculture (Maceljski 2002; Ivezić and Raspudić 1997). Maceljski (2002) estimates the annual loss at 6-25%, while Ivezić and Raspudić (2005) report average infestations of 50% during the past 10 years. Despite significant damage, the control of the ECB is carried out only in sweet and seed maize while potential losses in commercial maize are not addressed. Since sweet corn is meant for use in the fresh state or for canning/freezing, control of ECB is absolute. The pest can cause significant damage on sweet pepper (Maceljski, 2002), or some damage could occur occasionally on apples or potato. The damage on sweet pepper is mainly caused by second generation of moth which is not obligatory in all years and regions of Croatia (Maceljski, 2002).

The precondition for success in controlling the ECB is correctly estimating the time when the insecticide is to be applied. Insecticides are most effective when applied immediately after the emergence of the larvae from the eggs (Bartels and Hutchison 1995; Rinkleff et al. 1995). The date at which a given generation of ECB appears, and hence the dates for the application of the insecticide, are difficult to determine, for they depend on numerous climatic factors. The right time for the application of insecticide for the control of the ECB can be determined by the following methods: visual inspection of plants and determination of egg clusters and the number of recently hatched larvae, monitoring the flight of the ECB moths in a cage with last year's corn stalks, with an entomological lamp (Danon 1988), blacklight trap (Bartels and Hutchison 1995; Bartels et al. 1997) and with sexual pheromone traps (Bača 1976; Maceljski 1984; Bartels et al. 1997). Of all these methods, the best results are expected from pheromone traps.

First, pheromone for ECB (cis-11-tetradecenyl-acetate) was isolated by Klun & Brindley (1970). Klun & Junk (1977) published the isolation of 4 esters of pheromone for ECB, E-11-tetradecenyl acetate, Z-11-tetradecenyl acetate, E-9-tetradecenyl acetate and tetradecyl acetate. Two pheromone strains, E and Z, are most often in use in different countries. Since two strains are different in their attractiveness to given biotypes of ECB moth, it is necessary to determine the optimum ratios of cis- and trans- isomers that have the greatest attraction for a given population of ECB (Bača 1976; Bartels et al. 1997). Available on today's market are ready-made pheromone products with varying ratios of isomers. Pheromone products can be variously attractive to populations of ECB moths of different geographical origins. If there is no information about the composition of the biotypes in a corn borer moth population, each single pheromone product must be explored separately, in order to determine to what extent it can be employed in a particular area. In the case of ECB, pheromones are mainly used to determine the optimum insecticide application date. Pheromones are attractive to males. It is well known that in the case of ECB male moths complete their development earlier in the season. The question on the length of period between male moth appearance and larval hatching, i.e. need for insecticide application, arises.

In the period between 2002-2004 a study was performed in maize and in sweet pepper in order to determine: (1) which ECB pheromone biotype is prevalent in the two regions of Croatia, North West and North Croatia and which pheromone products were best suited for monitoring, (2) to determine the flight dynamics, and (3) to determine, with respect to flight dynamics, the most favorable timing of application of insecticides.

3.2. MATERIAL AND METHODS

The research was carried out during 2002-2004. In maize in 2002 and in 2004 the research was carried out at one site (Buzin in 2002 and Oborovo in 2004) and in 2003 at two sites (Buzin and Oborovo) located in North West Croatia. The distance between those two localities is approximately 20 km. In sweet pepper the research was carried out in 2002 at one site (Podravske

Sesvete) located in North Croatia, approximately 80 km far a way from Oborovo site.

The sweet maize hybrid "BC 276 su su" was sown at all experimental sites with maize. The sweet pepper variety "Istra F1" was planted on an experimental field. For monitoring of the first generation of ECB, pheromone lures produced by "Isagro", with the commercial names E, Z and E/Z were used. In each field, all three types of pheromones were evaluated using three replicates. Pheromone traps (plastic Delta-style trap, "Traptest", produced by Isagro) were placed in the maize fields on June 5, 2002, May 22, and June 3, 2003 and May 27, 2004. The same traps were placed in the sweet pepper field on May 31. Moths captured in the pheromone traps were inspected and removed every 2-4 days, up to mid July (1st generation). The total capture in the pheromone traps was analyzed by ANOVA (Vasilj, 2000) with means ranked according to the Student-Newman-Keuls ranking test (P=0.05).

In maize the percentage of plants with visible larval damage (feeding on leaves or hole on plant stalk) was determined by visual inspection on 100 plants in four replications two to three times each year. The visual inspection revealed that the percentage of damaged plants was made 20 and 35 days after the first ECB moth occurrence, i.e.19 and 34 days after the peak of the flight in 2002. In Buzin in 2003 the visual inspection was conducted 20 and 35 days after the first occurrence of moths i.e. 14 and 34 days after the peak of the flight. In Oborovo in 2003 the visual inspection was conducted 7 and 28 days after the first occurrence i.e. 5 and 26 days after the peak of flight. In Oborovo in 2004 the time period between the first occurrence and visual inspections was 30, 39 and 58 days and between the maximum flight and visual inspections 12, 23 and 49 days. The data on time period after the first occurrence and after the peak of the flight were correlated with the percent of infested plants established on the same date. All correlations were calculated by using ARM 7 software. Correlation coefficients, coefficients of determination and probabilities were calculated. Correlations were classified according Roemer-Orphal table (Vasilj, 2000). Regression equations were calculated by using Microsoft Excel.

3.3. RESULTS AND DISCUSSION

Generally, the moth captures in all experimental fields were low in all years and localities. Significantly highest catches obtained in maize in North West Croatia in both years were recorded using the E type pheromone lure

(Table 4). The catch with the E/Z type was significantly lower than the catch with E type and higher than the catch with the Z type of pheromone in 2002, while in 2003 no significant differences were observed between the catches obtained with E/Z and Z types of pheromones. Because of the very small number of ECB moths caught in maize in Oborovo in 2003 and 2004, it was not possible to make any valid conclusions about the differences among pheromone biotypes. In the sweet pepper field located in North Croatia significantly higher capture was noted on E/Z pheromone traps than on E or Z trap. The E type of trap caughts significantly higher amount of ECB moths than Z type as well. There is no information on the relative proportions of individual moth biotypes in Croatia. Bača (1976) states that the optimal ratio of pheromone isomers used to attract the male population of ECB at the Zemun polje site was 97:3 in favor of the Z pheromone. During his two year research, on this combination, 87.8% of males of the 1st generation and 71.8% of the male of the second generation were caught. Zemun polje is located far on the east and maybe those data could be compared with the data from the eastern part of Croatia. Our researches were carried out in North West and North Croatia. Based on results we can conclude that in North West Croatia, the E type is the predominant biotype in ECB population. The same was proved by Bažok et al. (2009) as well. In North Croatia the situation seems to be somewhat different, probably the population consists from both biotypes, but the future research should determine the ratio of the biotypes in the population.

Table 4. Mean (± SEM) total capture of ECB moths on various types of pheromone, Buzin, Croatia, 2002 and 2003

PHEROMONE PRODUCT	MAIZE		SWEET PEPPER
	BUZIN 2002	BUZIN 2003	PODRAVSKE SESVETE 2002
E	29.7 ± 11.22 a	15.0 ± 1.62 a	6.67 ± b
Z	2.3 ± 2.0 c	0.7 ± 0.66 b	2.67 ± c
E/Z	15.6 ± 1.76 b	3.3 ± 0.66 b	7.0 ± a
LSD* (P=0.05)	5.4	7.3	0.05

* Means followed by the same letter are not significantly different according to Student-Newman-Keuls test (P=0.05).

The pheromone traps were placed in the maize field in 2002 at the beginning of June, when according to most of the references the first ECB moth starts to appear. The first capture was recorded immediately after the traps were set, on June 4 and the maximum flight of moths was determined on June 6. In sweet pepper the traps were set up at the end of May and immediately after the traps exposure the maximum catch was recorded. The second maximum was recorded on June 25. These results (Figure 11) showed the need for the traps to be placed in the field as early as the end of May, which was subsequently performed in 2003 and 2004. Exceptionally high air temperatures in May 2003 led to earlier eclosion of the moths. The first capture in Buzin was recorded on May 24 and the maximum catch was recorded on May 30. The capture of the moth at the locality Oborovo in 2003 was very low, the first capture was recorded on June 6 and maximum captures two days later, on June 8. Later on the second maximum was recorded. In 2004, very low catches were recorded at the beginning of the flight. The first moths were captured on May 29 and first, but very weak maximum, was recorded on June 16. On July 6, the second maximum was recorded. The maximum moth catch determined in our investigation does not correlate with the findings by Hergula (1930), who stated that the maximum eclosion at two sites in Croatia was between July 8 and July 14. Other authors reported that the maximal appearance of the first generation of ECB moths is at the end of June and the beginning of July. The great percentage of plants on which moth damage was recorded at all localities already in June confirms the earlier occurrence of the moths than it was stated in the literature (Hergula, 1930; Schmidt, 1974; Maceljski, 2002). The earlier occurrence could be result of the different climatic condition in the year in which investigation is carried out or of the climate changing. According to Đulizibarić (1968) and Schmidt (1974) the occurrence of a given generation depends on temperature and humidity in given time periods in a given year. In Croatia ECB developed one and partially two generations per year (Maceljski, 2002). The observed changes in biology could lead to development of two generations of ECB per year. The research shows that the trap placement time had to be shifted forward to May 20-30.

When time periods after the first moth capture with the data on percent of infested plants were correlated (Figure 12), medium positive correlation (r=0.608) was established (P=0.0822).

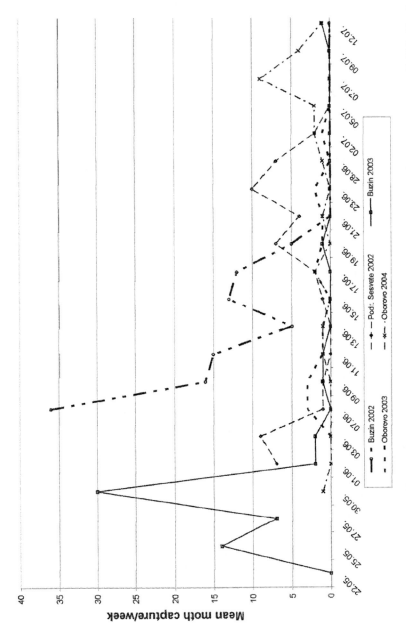

Figure11. Phenology of European corn borer flights in four maize fields in North West Croatia and one sweet pepper field in North Croatia.

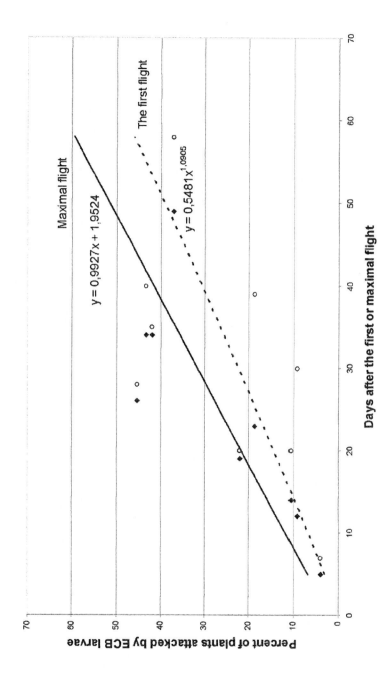

Figure 12. The correlations between the time periods after the first or the maximal ECB capture observed on pheromone traps and the percent of maize plants infested by larvae (the data from 4 trials).

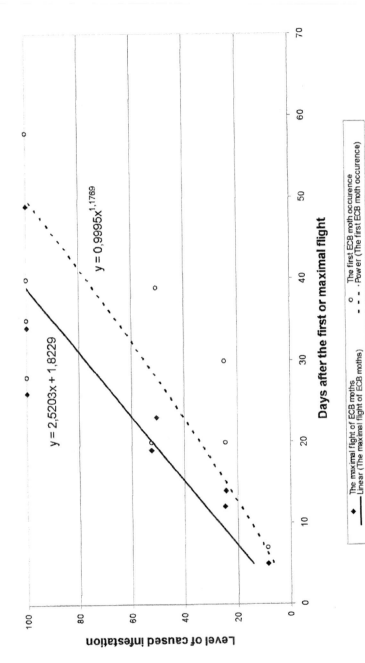

Figure 13. The correlations between the time periods after the first or the maximal ECB capture observed on pheromone traps and the percent of the observed damage (regarding to the total damage) on maize plants.

The amount of variability in the percent of infested plants by ECB larvae accounted for by the length of the period from first ECB moth capture as measured by the coefficient of determination (r^2) was quite low (0.37). Regression curve is power with the equation $y=0.5481x^{1.095}$. When time periods after the maximal moth capture with the data on percent of infested plants were correlated (Figure 11) high positive correlation (r=0.8244) was established (P=0.0063). The amount of variability in the percent of infested plants by ECB larvae accounted for by the length of the period from maximal ECB moth capture as measured by the coefficient of determination (r^2) was medium (0.6796). Regression curve is linear with the equation y=0.9927x + 1.9524. Since in the fields different population level was present, it could not be concluded on the level of infestation. Therefore the level of recorded damage at each observation was converted with regard to the maximal damage observed at the same locality. When the data on time periods after the first flight were correlated with those data (Figure 13), medium positive correlation (r=0.7035) was established (P= 0.0345). The amount of variability measured by coefficient of determination was medium (r^2= 0.495) and regression curve is power with the equation $y = 0.9995x^{1.1769}$. The same correlation was calculated for the time periods between maximal flight occurrence and level of damages and very strong positive correlation (r=0.8928) was established (P= 0.0012). The coefficient of determination (r^2) for those two variables was 0.7971. Regression curve is linear with the equation of y = 2.5203x + 1.8229. Since the correlations between the period from maximal ECB capture versus the damage recorded are stronger than correlation between the period from the first capture of moth on pheromone versus the damage recorded, it could be assumed that the peak of the flight should be used in determination of the most suitable time for insecticide application. Bažok et al. (2009) suggested that monitoring of the moths does not aim to determine the intensity of the infestation, but rather to demarcate the period of maximum moth incidence. Visual inspection of plants should commence at this time in order to determine the percent of infested plants or the number of egg clusters on plant leaves. The application of the insecticides according to Bažok et al. (2009) should be set to coincide with a delay of 10-14 days after the maximum catch. According to the obtained results (Figure 13), 10-14 days after the peak of the flight 25 to 40% of all damages would occur already.

Based on the ECB, moth capture the level of plants infestation could not be predicted. At the Oborovo site during 2003 and 2004, very few moths were captured by pheromone traps. Although very low captures on the trap occurred in Oborovo in 2003 and 2004 they resulted with maximum of 45.2 and 37.11

% of infested plants, respectively. It is possible that the low catches may have resulted from the location of the test plots near to the Sava river. According to some research, as well as the instructions of the manufacturers, the placement of pheromones close to large water areas should be avoided, as large water areas can affect the attractiveness of the pheromone bait. Generally, the number of ECB caught in all three years was very low, especially as compared with the capture in some US states (Bartels and Hutchinson, 1995). Despite considerable variability in the numbers of moths caught with pheromones at different sites throughout the years, the percentage of plants with damage in experimental fields did not differ substantially over the years or among the sites.

3.4. CONCLUSION

According to the conducted research, the predominant biotype of the ECB in North West Croatia is attracted by E type of pheromone, while in North Croatia the population is consists of both, E and Z biotypes. But due to low attractiveness of Z type, probably the majority of ECB population is still attracted by E biotype. Prior to conducting any research on the possible use of pheromones in determination of population level or in determination of proper insecticide timing, it is necessary to determine the optimum ratios of cis- and trans- isomers that have the greatest attraction for a given population of ECB. The differences in the respond of various ECB populations are possible although the distance between population is not very high. Therefore an organized mapping of ECB population within the country would help in future use of pheromones for ECB.

Due to the changes in climatic conditions which occurred in the past few years or due to extreme climatic condition of given year, the eclosion of the moths regularly occurs 15-30 days earlier than it was reported in older literature data. The reported changes in biology could lead to the development of two generations of ECB per year. These could even increase the harmfulness of ECB.

The most favorable timing of the application of insecticides should be determined regarding to the peak of the occurrence of ECB moths on pheromones. It should be set to coincide with a delay of maximum 10 days after the maximum catch because 10 days after the peak of the flight 25% of all damage would occur already.

REFERENCES

Bača, F. (1976). Ispitivanje sintetičkih feromona – seksualnih atraktanata – kukuruznog plamenca (*Ostrinia nubilalis* Hbn). *Z.bilja, 137/138*, 357-360. (in Serbian)

Bača, F. (1994). Novi član štetne entomofaune u Jugoslaviji *Diabrotica virgifera virgifera* LeConte (Coleoptera, Chrysomelidae). *Z. bilja, 45*, 125-131. (in Serbian)

Balarin, I. (1974). Fauna Heteroptera na krmnim leguminozama i prirodnim livadama u SR Hrvatskoj. *PhD Thesis, Poljoprivredni fakultet, Zagreb*. (In Croatian)

Barna, Gy; Edwards, CR; Gerber, C; Bledsoe, LW; Kiss J. (1998). Management of western corn rootworm (*Diabrotica virgifera virgifera* LeConte) in corn based on survey information from previous soybean crop. *Acta Phytopat. Entomol. Hung., 33*, 173-182.

Bartels, W.D. & Hutchison, D.W. (1995). On-Farm efficacy of aerially applied *Bacillus thuringiensis* for european corn borer (Lepidoptera: Pyralidae) and corn earworm (Lepidoptera: Noctuidae) control in sweet corn. *J. Econ. Entom. 88*, 380-386.

Bartels WD; Hutchinson DW; Udayagirl, S. (1997). Pheromone trap monitoring of Z-strain European corn borer (Lepidoptera: Pyralidae): optimum pheromone blend, comparasion with blachlight traps, and trap number requirements. *J. Econ. Entom. 90*, 449-457.

Bartelt, R.J. & Chiang, H.C. (1977). Field Studies Involving the Sex-attractant Pheromones of the Western and Northern Corn Rootworm Beetles. *Environ. Entomol., 6*, 853-861.

Bažok, R; Igrc Barčić, J; Kos, T; Gotlin Čuljak, T; Šilović, M; Jelovčan, S; Kozina A. (2009). Monitoring and efficacy of selected insecticides for

European corn borer (Ostrinia nubilalis Hubn., Lepidoptera: Crambidae) control. *J.Pest Sci.,82(3)*. Doi: 10.1007/s10340-009-0255-z

Berger, H.K. (1996). Multicountry coordination efforts to deal with the western corn rootworm (*Diabrotica virgifera virgifera*). *IWGO News Letter 16(1)*, 26-29.

Bertrossa, M; Derron, J; Brunetti, LC. (2001). Update of monitoring data of *Diabrotica virgifera virgifera* LeConte in Switzerland in 2001. *Proceedings of XXI IWGO Conference,* Legnaro, Italy, November 2001, 169-173.

Blackshaw, R.P. & Vernon, R.S. (2008): Spatial relationships between two *Agriotes* click-beetle species and wireworms in agricultural fields. *Agric.For. Entomol. 10,* 1-11.

Blomquist, GJ; Jurenka, R; Schal, C; Tittiger, C. Biochemistry and Molecular Biology of Pheromone Production. In: Gilbert, LI; Iatrou, K; Gill, S.S. (Eds). *Molecular Insecrt Science* Oxford: Elsevier; 2005; *Vol. 3,* 705-751.

Borg-Karlson, AK; Agren, L; Dobson, H; Bergstrom, G. (1988). Identification and electroantennographic activity of seks-specific geranyl esters in a abdominal gland of female *Agriotes obscurus* (L.) and *A. lineatus* (L.) (Coleoptera: Elateridae). *Experientia 44,* 531-534.

Butenandt, A; Beckmann, R; Stamm, D; Hecker, E. (1959). Über dem sexual-lockstoff des seidenspinners *Bombyx mori:* Reindarstellung und konstitution. *Z. Naturforsch. A 14,* 283-284.

Chabert, A; Blot, Y. (1992) Estimation des populations larvaires de taupins par un piege attractif. *Phytoma, 436,* 26-30.

Ciosi, M; Miller, NJ; Kim, KS; Giordanno, R; Estoup, A; Guillemaud, T. (2009). Invasion of Europe by the western corn rootworm: multiple transatlantic introductions with various reductions of genetic diversity. *23rd IWGO Conference & 2nd International Conference of Diabrotica Genetics, Munich, Germany, Book of Abstract, Scientific session 6,* 28-29, (http://www.iwgo.org/)

Cuperus, GW; Mulder, PG; Royer, TJ. Implementation of Ecologically-Based IPM. In: Rechcigl, JE; Rechcigl, NA. (eds). *Integrated Pest Management Techniques for Environmental Protection.* Boca Raton: CRC Press, 2000; 171-204.

Čamprag, D. *Skočibube i integralne mere suzbijanja.* Poljoprivredni fakultet, Institut za zaštitu bilja, dr. Pavle Vukasović, Novi Sad, 1997. (In Serbian)

Danon, V. (1988). Kukuruzni moljac *Ostrinia nubilalis* Hb i mogućnosti suzbijanja. *Poljoprivredne aktualnosti, 1/2,* 161-172. (In Croatian)

Dent, D. *Insect Pest Management.* CABI Publishong; Cambridge; 2000.

Conway, GR. Tactical models. In: Conway, GR. (ed). *Pest and Pathogen Control: Strategic, Tactical and Policy Models.* John Wiley and Sons; Chichester; 1984; 15-28.

Dobrinčić, R. (2001). Istraživanje biologije i ekologije *Diabrotica virgifera virgifera* LeConte – novog člana entomofaune Hrvatske. *PhD Thesis, Agronomski fakultet, Zagreb.* (In Croatian)

Đulizibarić, T. (1968). Rezultati ogleda kod suzbijanja kukuruznog plamenca (*Ostrinia nubilalis* Hb.) insekticidima. *Biljni lekar 3/4,* 33-37. (In Serbian)

Edwards, CR; Barna, G; Kiss, J. (1997). Trap catch comparisons between Hungarian pheromone and Pherocon AM Traps. *IWGO- News Letter 18,* 28-29.

Ester, A; van Rozen, K; Griepink, FC. (2001). Previous research of monitoring of *Agriotes spp.* with sex pheromones. *Proceedings of XXI IWGO Conference,* Legnaro, Italy, November 2001, 305-310.

Furlan, L; Toth, M; Ujvary, I; Toffanin, F. (1996). L'utilizzo di feromoni sessuali per la razionalizzazione dela lotta agli elateridi del genere Agriotes: prime sperimentazioni in Italia. *ATTI Giornate Fitopatologiche,1,* 133-140.

Furlan, L; Toth, M; Ujvary, I. (1997). The suitability of sex pheromone traps for implementing IPM strategies against Agriotes populations (Coleoptera: Elateridae). *Proceedings of XIX IWGO Conference,* Guimares, August 30- September 5, 173-182.

Furlan, L; Vettorazzo, M; Ortez, A; Frausin, C. (1998). Western corn rootworm (*Diabrotica virgifera virgifera* LeConte) presence in Italy. *IWGO News Letter 18(2),* 15.

Furlan, L; Toth, M. & COOPERATORS (1999). Evaluation of the effectiveness of the new Agriotes sex pheromone traps in different European countries. *Proceedings of XX IWGO Conference,* Adana, Turkey, 4-10 September, 171-175.

Furlan, L; Vettorazzo, M; Frausin, C. (1999a). Effectiveness of containment strategies against *Diabrotica virgifera virgifera* LeConte populations in north eastern Italy. *4th FAO/TCP Meeting, 5th EPPO ad hoc Panel and 6th International IWGO Workshop, Summary of Abstracts,* Paris, France, 40-41.

Furlan, L; Di Bernardo, A; Maini, S; Ferrari, R; Boriani, M; Nobili, P; Bourlot, G; Turchi, A; Vacante, V; Bonsignore, C; Giglioli, G; Toth, M. (2001). First practical results of click beetle trapping with pheromone traps in Italy. *Proceedings of XXI IWGO Conference,* Legnaro, Italy, November, 277-282.

Furlan, L; Toth, M; Parker,WE; Ivezić, M; Pančić, S; Brmež, M; Dobrinčić, R; Barčić, J; Muresan, F; Subchev, M; Toshova, T; Molnar, Z; Ditsch, B; Voigt, D. (2001a). The efficacy of the new Agriotes sex pheromone traps in detecting wireworm population levels in different European countries. *Proceedings of IWGO Conference* Legnaro, Italy, November, 283-316.

Furlan, L; Di Bernardo, A; Girolami, V; Vettorazzo, M; Piccolo, AM; Santamaria, G; Donatoni, L; Funes, V. (2001b). *Diabrotica virgifera virgifera* eradication-cantainment temptative in Veneto Region: year 2001. *Proceedings of IWGO Conference* Legnaro, Italy, November, 47-52.

Furlan, L. (2009). Can we produce maize without soil and seed treatment insecticides? *23rd IWGO Conference & 2nd International Conference of Diabrotica Genetics, Munich, Germany, Book of Abstract, Scientific session 11.* 49-49 (http://www.iwgo.org/)

Furlan, F; Burgio, G; De Luigi, V; Palmieri, S; Vettorazzo, M; Zanini, G. (2009). The *Diabrotica virgifera virgifera* eradication in Venice focus area has been accomplished. *23rd IWGO Conference & 2nd International Conference of Diabrotica Genetics, Munich, Germany, Book of Abstract, Scientific session 11.* 14-15. (http://www.iwgo.org/)

Gomboc, S; Milevoj, L; Furlan, L; Toth, M; Bitenc, P; Bobnar, A; Celar, F. (2001). Two-years monitoring Click Beetle and Wireworms in Slovenia. *Proceedings of XXI IWGO Conference*, Legnaro, Italy, November, 283-292.

Guss, PL. (1976). The Sex Pheromone of the Western Corn Rootworm (*Diabrotica virgifera*). *Environ. Entomol., 5,* 219-223.

Guss, PL; Tumlinson, JH; Sonnet, PE; Proveaux, AT. (1981). Identification of a Female –Produced Sex Pheromone of the Western Corn Rootworm. *J. Chem. Ecol., 8,* 545-556.

Guss, PL; Sonnet, PE; Carney, RL; Branson, TF; Tumlinson, JH. (1984). Response of *Diabrotica virgifera virgifera, D.v. zeae* and *D. porracea* to stereoisomers of 8-methyl-2-decyl propanoate. *J. Chem. Ecol., 11,* 21-26.

Hein, G.L. & Tollefson, J.J. (1985). Use of the Pherocon AM Trap as a Scouting Tool for Predicting Damage by Corn Rootworm (Coleoptera: Chrysomelidae) Larvae, *J. Econ. Entomol., 78,* 200-203.

Hergula, B. (1930). Daljnji prilog proučavanja kukuruzova moljca (*Pyrausta nubilalis* Hubner) i njegovih parazita u Jugoslaviji. *Glasnik Jugoslovenskog Entomološkog društva V-VI (1-2),* 98-117.

Hicks, H. & Blackshaw, R.P. (2008). Differential responses of three *Agriotes* click beetle species to pheromone traps. *Agric.For. Entomol. 10,* 443-448.

Igrc Barčić, J. (1996). First results of comparative investigations of the atractiveness of various baits to the WCR. *IWGO News letter 16(2)*, 22-23.

Igrc Barčić, J. & Maceljski, M. (1996). Monitoring *Diabrotica virgifera virgifera* LeConte in Croatia in 1995. *IWGO News letter, 16(1)*, 11-13.

Igrc Barčić, J. & Dobrinčić, R. (1998). Results of monitoring *Diabrotica virgifera virgifera* LeConte in 1998 in Croatia, Abstract, IWGO News letter 18(2), 9.

Igrc Barčić, J; Dobrinčić, R; Maceljski, M. (1999). The spread and population density of the western corn rootworm (*Diabrotica virgifera virgifera*) in Croatia in 1999. *4th FAO/TCP Meeting, 5th EPPO ad hoc Panel and 6th International IWGO Workshop, Summary of Abstracts*, Paris, France, 11-12.

Igrc Barčić, J. & Bažok, R. (2004). The influence of different food sources on the life parameters of western corn rootworm (*Diabrotica virgifera virgifera* LeConte, Coleoptera: Chrysomelidae). *Razprave IV. Razreda SAZU, XLV-1*, 75-85.

Ilovai, Z. (1996). Monitoring of western corn rootworm (*Diabrotica virgifera virgifera* LeConte) in Hungary in 1996. *IWGO News Letter 16(2)*, 18.

Ivezić, M; Raspudić, E. (1997). Intensity of attack of the corn borer (*Ostrinia nubilalis* Hűbner) on the teritory of Baranja in the preiod 1971-1990. *Nat Cro 6*, 137-142.

Ivezić, M; Raspudić, E. (2005). Ekonomski značajni štetnici kukuruza na području istočne Hrvatske. *Razprave IV. Razreda SAZU, Dissertationes, XLV-1*, 87-98. (In Croatian)

Karabatsas, K; Tsakiris, V; Zarpas, K; Tsisipis, JA; Furlan, L; Toth, M. (2001).

Seasonal fluctuation of adult and larvae *Agriotes spp.* (Coleoptera: Elateridae) in Central Greece. *Proceedings of XXI IWGO Conference*, Legnaro, Italy, November, 269-276.

Kereši, TB; Čamprag, D; Sekulić, RR. (1997). Comparison of trapping methods for monitoring of *Diabrotica virgifera virgifera* LeConte adults. *IWGO- News Letter 17*, 13-14.

Kiss, J; Edwards, RC; Berger, HK; Cate, P; Cean, M; Cheek, S; Derron, J; Festić, H; Furlan, L; Igrc Barčić, J; Ivanova, I; Lammers, W; Omelyuta, V; Prinzinger, G; Reynaud, P; Sivčev, I; Siviček, P; Urek, G; Vahala, O. Monitoring of Western Corn Rootworm (*Diabrotica virgifera virgifera* LeConte) in Europe 1992-2003. In: Vidal, S; Kuhlman, U; Edwards, CR.

(Eds). *Western Corn Rootworm Ecology and Management.* Oxford: CABI Publishers; 2005; 121-144.

Klun, J.A. & Brindley, T.A. (1970). Cis-11-tetradecenyl acetate, a sex stimulant of the European corn borer. *J. Econ. Entomol., 63,* 779-780.

Klun, J.A. & Junk, G.A. (1977). Iowa European corn borer sex pheromone. Isolation and identification of four C14 esters. *J. Chem. Ecol.,3,* 447-459.

Kogan, M. (1998). Integrated Pest Management: Historical perspectives and contemporary developments. *Ann. Rev. Entomol. 43,* 243-270.

Kovačević, Ž. (1960). Problematika zemljišnih štetnika u istočnoj Slavoniji. *Sav. Polj. 7/8,* 567-580.

Kozina, A; Igrc Barčić, J; Bažok, R; Kos, T. (2009). Distribution and abundance of *Agriotes ustulatus* in three regions in Croatia. *23rd IWGO Conference & 2nd International Conference of Diabrotica Genetics, Munich, Germany, Book of Abstract,* Poster 17

Kudryavstev, I; Siirde, K; Lääts, K; Ismailov, V; Pristavko, V. (1993). Determination of distribution of harmful click beetle species (Coleoptera, Elateridae) by syntetic sex pheromones. *J. Chem. Ecol., 19(8),* 1607-1611.

Lance, D.R. (1988). Potential of 8-methyl-2-decyl Propanoate and Plant Derived Volatiles for Attracting Corn Rootworm Beetles (Coleoptera: Chrysomelidae) to Toxic Bait. *J. Econ. Entomol., 81,* 1359-1362.

Maceljski, M. (1975). Iskustva i rezultati višegodišnjih ispitivanja suzbijanja žičnjaka u kukuruzu. *Agr. Glas. 1 (4),* 127-141. (In Croatian)

Maceljski, M. (1984). Biotehnički insekticidi s posebnim osvrtom na feromone. *Polj znan sm 67,* 609-618. (In Croatian)

Maceljski, M. *Poljoprivredna entomologija.* Čakovec: Zrinski; 2002. (In Croatian)

Maceljski, M. & Bedeković, M. (1962). Novi momenti zaštite kukuruza od štetnika. *PF Zag. Sav. Z.bilja,* 151-160. (In Croatian)

Metcalf, R.L. *Foreword.* In: Krysan, J.L. & Miller, T.A. (Eds). *Methods for the Study of Pest Diabrotica.* New York: Springer Verlag, VII-XV.

Novak, P. *Kornjaši jadranskog primorja.* Zagreb: JAZU; (1952). (In Croatian)

Prinzinger, G. (1996). Monitoring of western corn rootworm (*Diabrotica virgifera virgifera* LeConte) in Hungary in 1995. *IWGO News Letter 16(1),* 7-11.

Reynaud, P. (2002). First occurrence of *Diabrotica virgifera virgifera* in France. *IWGO News Letter 23(2),* 20-21.

Rinkleff, JH; Hutchinson, WD; Campbell CD; Bolin PC; Bartels DW. (1995). Insecticide toxicity in european corn borer (Lepidoptera Pyralidae):

ovicidal activity and residual mortality to neonates. *J Econ Entom, 88*, 246-253.

Statistical Yearbook of the Republic of Croatia, Državni zavod za statistiku; Zagreb; 1996; 1997; 1998; 1999; 2000; 2001; 2002; 2003; 2004; 2005; 2006.

Suckling, D.M. & Karg, G. Pheromones and Other Semiochemicals. In: Rechcigl, J.E., Rechcigl, N.A. (eds.), *Biological and Biotechnological Control of Insect Pests*, Boca Raton: CRC Press; 2000; 63-99.

Schmidt, L. (1974). Rezultati praćenja leta kukuruznog moljca (*Ostrinia nubilalis* Hb.) na poljoprivredno-prehrambenom kombinatu Županja. *Biljna zaštita*, 66-68. (In Croatian)

Štrbac, P. (1983). Fauna, bionomija i morfološko-taksonomske karakteristike klisnjaka i trčuljka (Col.: Elateridae; Carabidae) u agroekološkim uslovima Slavonije i Baranje. *PhD Thesis, Poljoprivredni faluktet Osijek.* (In Croatian)

Toth, M; Toth, V; Ujvary, I; Sivčev, I; Manojlović, B; Ilovai, Z. (1996). Sex pheromone trapping of *Diabrotica virgifera virgifera* LeConte in Central Europe. *Nővényvédelem 32(9)*, 447-452.

Toth, M; Furlan, L; Szarukan, I; Ujvary, I; Yatsynin, VG; Sivčev, I; Ilovai, Z. (1997). Development of new pheromone traps for beetle pests (Coleoptera) of maize. *IWGO- News Letter, 17*, 11.

Toth, M; Furlan, L; Szarukan, I; Ujvary, I; Yatsynin, VG; Tolasch, T; Francke, W. (1998). Development of pheromone traps for European click beetle pests (Coleoptera: Elateridae). *2nd International Symposium on Insect Pheromones, WICC- International Agricultural Centre Wagenigen, The Netherlands, 30 March- 3 April 1998*, 53.

Toth, M; Imrei, Z; Sivčev, I; Tomašev, I. (2000). Recent advances in trapping methods of *Diabrotica v. virgifera* : high-capacity, non-sticky traps and effective trapping range. *IWGO- News Letter 22*, 31-32.

Toth, M; Furlan, L; Szarukan, I; Ujvary, I. (2001). Geranyl hexanoate attracting males of click beetles *Agriotes rufipalpis* Brulle and *A. sordidus* Illiger (Coleoptera: Chrysomelidae) *J. Appl. Entomol. 126*, 312-314.

Toth, M; Imrei, Z; Szarukan, I; Kőrősi, R; Furlan, L. (2001a). First results of click beetle trapping with pheromone traps in Hungary 1998-2000. *Proceedings of XXI IWGO Conference*, Legnaro, Italy, November, 263-267.

Toth, M; Furlan, L; Szarukan, I; Yatsynin, VG; Ujvary, I; Tolasch, T; Francke W; Imrei, Z; Subchev, M. (2002). Identification of the sex pheromone composition of the click beetle *Agriotes brevis* Candeze (Coleoptera: Chrysomelidae). *J. Chem. Ecol. 28,* 1641-1652.

Topfer, S. & Kuhlman, U. Natural Mortality Factors Acting on Western Corn Rootworm Populations: a Comparison between the United Stated and Central Europe. In: Vidal, S; Kuhlman, U; Edwards, CR; (eds.). *Western Corn Rootworm Ecology and Management,* Oxford: CABI Publishers; 2005; 95-120.

Vasilj, Đ. *Biometrika i eksperimentiranje u bilinogojstvu.* Zagreb: Hrvatsko agronomsko društvo; (2000). (In Croatian)

Wilde, G.E. (1999). Areawide management of corn rootworm in Kansas. *4th FAO/TCP Meeting, 5th EPPO ad hoc Panel and 6th International IWGO Workshop, Summary of Abstracts*, Paris, France, 26-27

Yatsynin, VG; Oleshchenko, IN; Rubanova, EV; Ismailov, VA. (1980). Identification of active components of sexual pheromones of *Agriotes gurgistanus, Agriotes litigiosus* and *Agriotes lineatus* click beetles. *Khim. Selsk. Khoz.* 33-35. (In Russian)

Yatsynin, VG; Karpenko, NN; Orlov, VN. (1986). Sex pheromone of the click beetle *Agriotes sputator* (Coleoptera, Elateridae). Khim. Komm. Zhivot., Edition Moskva, Nauka 53-57. (In Russian)

Yatsynin, VG; Rubanova, EV; Okhrimenko, NV. (1996). Identification of female-produced sex pheromones and their geographical difference in pheromone glands extract composition from click beetles (Col., Elateridae). *J. Appl. Entom. 120,* 463-466.

Yatsynin, G.V. & Rubanova, E.V. (2001). Objectives of the research on click beetle species in Kuban region. *Proceedings of XXI IWGO Conference*, Legnaro, Italy, November, 311-316.

Youngman, RR; Kuhar, T.P; Midgarden, DG. (1996). Effect of Trap Size on Efficiency of Yellow Sticky Traps for Sampling Western Corn Rootworm (Coleoptera: Chrysomelidae) Adults in Corn. *J. Entomol. Sci. 31(3),* 277-285.

Zseller, IH; Szell, E. (2000). Results of biological observations of Western corn rootworm in 2000 in Hungary. WCR, *Diabrotica virgifera virgifera* LeConte, *IWGO News letter, 22,* 22.

URL pages:

http://www.eppo.org/QUARANTINE/Diabrotica_virgifera/diabrotica_virgifera.htm

http://faostat.fao.org/

http://www.iwgo.org/

http://www.dzs.hr/Hrv/publication/stat_year.htm

INDEX